MW01533229

SERIES EDITORS

Dr. Sally Monsour
Georgia State University Atlanta

Mary Jarman Nelson
Music Consultant Winter Park, Florida

Sound Exploration and Discovery

by Mary Palmer

Assistant Professor of Music Education Florida Technological University Orlando

Classroom Music Enrichment Units

THE CENTER FOR APPLIED RESEARCH IN EDUCATION, INC.
521 FIFTH AVENUE, NEW YORK, N.Y. 10017

Library of Congress Cataloging in Publication Data

Palmer, Mary.
    Sound exploration and discovery.

    (Classroom music enrichment units)
    "Sources for further exploration": p.
    1. School music--Instruction and study. 2. Sound.
I. Title. II. Series.
MT10.P185S7   372.8'7'044   74-7237
ISBN 0-87628-217-6

Printed in the United States of America

About This Handbook

This handbook is a practical guide for teachers who wish to help children in their exploration and discovery of sound. Children should have many experiences with sound and its organization while they are in elementary school. These experiences will add to their musical sensitivity and, hopefully, lead them to further, independent exploration of sound.

Sounds can be symbolized in a variety of ways. Several of these are included here, though emphasis is on the sounds themselves rather than on their symbols. Nontraditional ways of notating sounds are presented, and children's sound notations are shown also.

The handbook provides many suggestions for the discovery of sound sources, and specific ideas, techniques, and sample lessons for getting children involved with sound. Children's own compositions are are encouraged throughout. The teacher's role is pictured as being the organizer of learning activities. He or she has the responsibility for establishing the environment and making suggestions to guide the child's experiences. He or she becomes a facilitator for learning rather than a dispenser of knowledge.

The ideas presented here can be used as a springboard to creative teaching. They represent a beginning point for the exploration and discovery of sound. Teachers are encouraged to build on these suggestions in order to meet the individual needs of the children they teach.

Without extensive musical training classroom teachers should be comfortable using this handbook because its nontechnical language enables them to experiment with sound exploration lessons confidently. Teachers with a musical background will be helped to nurture creativity and musical composition in their students.

Mary Palmer

Contents

1

Environmental Sounds

Sounds of the environment frequently go unnoticed in our busy world. Yet, when we listen to them, everyday sounds become interesting and even exciting. This section includes excerpts from actual lessons and suggestions on how you will be able to explore environmental sounds in your classroom.

The raw material of music is *sound*. There are many environmental sound sources that are available for use in the music classroom. When children are allowed to explore sound, they become active participants in the discovery of sound possibilities. They become *involved* with sound exploration.

Emphasizing Everyday Sounds

Zoooom! "That was a Datsun 240Z, Joe."

Varoooom! "Yeah, Tom. And *that* was a Corvette." (And they weren't even *looking!*)

Are these the same two boys who can't seem to hear anything in your classroom? Why can they make these fine discriminations between sounds when they can't find page 97 in their math books upon your request? Probably the boys were truly *listening* to the sounds produced by the passing cars, whereas they were not listening to you. Perhaps we can capitalize on children's powers of auditory perception by encouraging them to be sensitive to many everyday sounds.

What is the first sound that you hear in the morning?
What is your favorite sound?
What sounds do you hear when you take a ride into the city?
What sounds are around when you ride your bicycle by the lake?

The worldwide noise level is increasing by one decibel every year. Consequently, our interest in sound could have environmental significance—we may help to stamp out noise pollution. Hugh Downs has indicated the possible need for "earlids." Since earlids don't seem to be forthcoming, the need for discriminating among sounds is imperative.

Ask your class to listen without making any sounds themselves for thirty seconds. What did they hear?

—the breathing of their classmates
—the clock ticking
—the lights buzzing
—someone walking down the hall
—Joe sneezing
—the shuffling of feet
—the ringing of a distant bell

To continue the exploration of sounds, you might ask the children to gather sound-producing objects from home. Ask them to explore the garage, the kitchen, the boathouse, and the backyard for sound producers.

You might perform several sounds for your class and have them identify the source of the sound. Ask the children to close their eyes so that they can concentrate with their ears only. The following list of ideas may be helpful:

Drop a plastic bracelet
Shake a plastic container in which you have placed rice
Toss a baseball into the metal wastebasket
Tap a water glass with a spoon
Strike a metal rod with another metal rod

Young children may simply identify the sound producers. Older children may begin to think about the "hows" and "whys."

—How did you know that the object I dropped was plastic? (Because it seemed to bounce.)
—How did you know that I placed rice and not stones in the plastic container? (Because it sounded "light.")
—How could you tell that I tossed a hard baseball into the wastebasket? (Because it landed with a thud and the sound was resonated.)
—How did you know that I hit a metal object with another metal object? Why didn't you think that I hit a wooden board with a metal rod? (Because the sound reverberated.)

Causes of Sound

This experience can be developed into an exploration of the causes of sound. What is necessary for a sound to be produced? There must be something or someone to start the sound. Vibrations must occur. A resonating chamber brings out the sound. Frequently, production of sound requires some kind of manipulation. You could drop something in order to make a sound; you could pluck something in order to make a sound; you could strike one object with another in order to make a sound. Ask your class to make a list on the chalkboard of ways that they have produced sounds.

Developing Your Own Environmental Instruments

Ask children to devise instruments on which they can produce sounds in a variety of ways. Jim places rubber bands over an empty box cover. With this devised instrument, he can produce a sound by plucking the rubber bands with his fingers, by tapping on the side of the box with his pencil, and by sliding a rubber mallet over the bands he produces yet another sound. By encouraging children to explore unconventional ways of using instruments of their own creation, as well as conventional classroom instruments, you aid them in developing a greater vocabulary of sound possibilities.

When your class has assembled an interesting collection of sound-producers, encourage the children to explore all of the sound possibilities that they have discovered. Perhaps the class could classify the sounds into appropriate categories that might include metallic sounds, scraping sounds, harsh sounds, high sounds, and low sounds. One class found these sound-producers:

Nonpitched:
—metal wastebaskets (Play by striking the outside with a pencil; strike the inside. Is the sound the same?)
—paper bags (Try crumpling the bag to produce a sound; blow it up with air and pop the bag.)
—a mobile made of shells
—grating carrots (The sound of grating carrots is interesting; scrape a pencil up and down on the grater to produce a different sound.)
—a typewriter

—pot pie pans taped together with popcorn kernels inside

—a coffee can with a balloon stretched over the top

—Styrofoam cups taped together with rice inside

—sewer piping with plastic covers—place popcorn kernels inside (This student's subdivision was just putting in a public sewer system and sewer piping was readily available!)

—salt and pepper shakers

—egg cartons

Pitched:

—xylophone made of varying lengths of boards that have been tied together (Try playing this with it lying on the floor; then try playing it when it is hanging up. Does it sound different?)

—different lengths of bamboo hanging from a vertical bar

—different-sized covered plastic containers (à la tympani drums)

—empty tin cans of various sizes (Play them when they are turned upside down; turn them over and fill with varying amounts of water. Cover the cans with plastic wrap and secure it with a rubber band. One student discovered that this method of playing the cans produced a very different sound.)

—different-sized nails, pounded into a board; strike with a metal rod (Consider the pitches produced.)

—pairs of nails, pounded into a board—pairs should be separated by different-sized spaces. Place rubber bands between the pairs of nails.

—plastic egg carton with a hole cut into the top. Place a string from one end of the carton to the other. Put a pencil under the string. By moving the pencil around and then plucking the string, one can produce different pitches.

This experience of listening carefully to a great variety of sounds could lead to a discussion attempting to define the question, "What is *music?*" Direct the discussion to include the following points:

Music is *pitches*—high and low

Music is *dynamics*—loud and soft

Music is *tempo*—fast and slow

Music is *textures*—thick and thin

Music is *sound* that is organized

When children are having exciting new experiences with sound, the development of music vocabulary is natural. The use of musical terms within the context of the musical experience provides the child with concrete examples of the new terms. In aiding your class with its vocabulary development, you may wish to use new terms in a musical context several times before defining them. When the children have become familiar with a new term, write it on the chalkboard for them to see.

Jimmy played his various-sized tin cans. The children noticed that each can had a different sound. Their teacher told them that each can produced a different *pitch*. When Susan played her wooden xylophone, the class heard different pitches. After the class had experienced pitch with a variety of different instruments, the teacher asked for a definition of the term "pitch." The children determined that pitch refers to the highness or lowness of a sound. Further, they discovered that the high pitches are produced by small objects, such as a tuna fish can or a short xylophone board. Larger cans and longer boards produced lower pitches. This discovery led the children to examine the box of resonator bells (each bell is an individual block with its own resonator) to see if the larger bells had a lower pitch than the smaller ones. Of course, they discovered that this same principle was true for the resonator bells. Later in the week, the children began to bring various-sized wastebaskets, bottles, and nails to show that these, too, produced different pitches. These children *experienced* pitch within a musical context. Over a period of several weeks, they developed a concrete understanding of "pitch."

Organizing "Found Sounds"

One way to classify "found sounds" is by the *way* in which they are played. Sounds can be played by:

—striking
—scraping
—rattling
—shaking
—dropping
—plucking
—blowing

Can your class think of other ways?

Another way to classify these sounds is by their tone color.[1]

—harsh
—sharp
—smooth
—delicate

"When two sounds are played together do they blend? Does one stand out?" These are important decisions when exploring possible ways of combining sounds.

Developing a Classroom Orchestra

After classifying the sounds, you will probably want to develop a classroom orchestra made up of your "found sounds." Have all children producing sounds by striking their "instruments" seated together, all those creating sounds by scraping objects seated together, and so forth. Elect a conductor who will lead the improvised performance. Explore the effects of tempo and dynamic contrasts, of solo performances by one instrument in a section versus all of the instruments in that section, of the combined sound produced by two sections versus two different sections combined. When the total "orchestra" plays together, which section stands out? Which section isn't heard at all? To help the children answer these questions, make a tape recording of their improvisation so that they can concentrate on the overall sound.

When the children have discovered that a particular type of instrument can't be heard when the total orchestra is playing, they will discover the need to feature that type in a solo passage. This focused listening may lead the children to discover the dynamic capabilities of each instrument, as well as its pitch range. What pitch range is the most easily heard when the total "orchestra" plays together?

One class discovered that their *metal* instruments were the most resonant. Their classroom orchestra composition was an "additive" one in which the plucked instruments began by playing an ostinato

---

[1] Musicians use the word *timbre* to denote the tone color or unique sound of an instrument. The timbre of a flute allows us to distinguish its sound from that of a trumpet.

pattern that group had created. Then the plastic instruments filled with various materials were added to the first ostinato. Then the wooden instruments were added to the first two patterns. Finally, the metal instruments were playing. By this time, four sections of instruments were playing. Naturally, this resulted in a gradual crescendo of sound. When the dynamic high point was reached, each group, one by one, stopped playing its part. In this composition, each "section" of instruments was easily heard because of the way in which they were added to the composition. Their composition looked like the pattern in Figure 1-1.

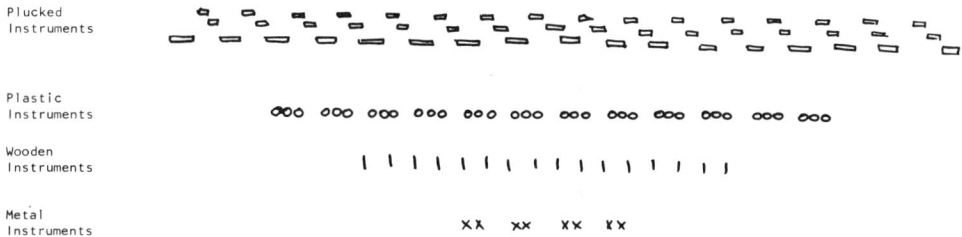

Plucked
Instruments

Plastic
Instruments

Wooden
Instruments

Metal
Instruments

Figure 1-1

Sample Experience with Environmental Sounds

Objective: To focus exploration on the variety of sound and sound-producing techniques available using kitchen utensils

Materials: spoons, forks, knives, egg beater, drainer, strainer, pie tin, jar with rice in it, plastic cup with popcorn kernels in it, any other kitchen utensils available

Procedures:
1. Distribute the kitchen utensils to students in the class.
2. Provide time for the students to explore the sound possibilities of their "instruments." Ask the students to discover *three different* sounds using whatever utensil has been given to them.
3. Ask the children to close their eyes as they listen to other children play their "instruments." Discuss what is being heard. Can they identify the *producer* (i.e., which utensil) of the sounds heard? Can they discover *how* the sounds are produced (i.e., scrape, hit, drop, etc.) on that particular instrument?

2

Imitating Sounds

Poetry can be enhanced when sound effects are added to the spoken words. For example, sounds can simulate moods from poetry and recreate events. Images come to life when they are given sound. Photographs live when sound accompanies them.

The study of science can be enlivened when children explore mechanical devices such as the tape recorder. The tape recorder provides for "self-discovery" as well as exploration and manipulation of environmental sounds.

This section suggests ways of using sound effects with poetry, imagery, and photography. Suggestions on the use of the tape recorder in the elementary classroom are also explored.

Sound Effects

The discovery of sound effects to enrich a song, story, poem, or dramatization can enhance a child's perception of the sound that is to be reproduced. Sound effects are reproductions of actual sounds heard around us. Perhaps work with sound effects should be considered to be imitative rather than creative work. Children may explore vocal, body, and instrumental sounds in their attempts to produce satisfying reproductions of familiar sounds.

Halloween is a very exciting time for young children. Many sounds are associated with Halloween. Therefore, this holiday may provide a fertile beginning for the exploration of sound effects. One teacher had her class make a list of all of the sounds that they thought of in relation to Halloween. The following list shows some of their ideas:

—witches flying on broomsticks
—ghosts crying
—bats flying

—black cats screeching
—wind whirring
—creaking doors in the haunted house

The children used sandblocks to represent the flying witches. The crying ghosts were reproduced by scratching fingers over the head of a hand drum. A high-pitched vocal screech imitated the sound made by bats; quick flaps of the fingers into the palms of the hands represented the flapping of the bats' wings. The wind was imitated with vocal sounds. The creaking doors were played by a slow stroke over the notched woodblock. All of these sounds created a Halloweenlike mood. The song "On Halloween," [1] or the poem "Black and Gold" [2] by Nancy Byrd Turner would be appropriate stimuli for the creation of sound effects.

Creating a Mood Environment

Poetry embodies many characteristics of music: rhythm, pulse, tempo, tone color, and expression. The study of poetry can contribute to children's sensitivity to verbal expression. By discovering appropriate instrumental accompaniments to the poetry, children can become more sensitive to the sound characteristics of the words.

"Gargoyle!" by Kay Maves provides an interesting study in contrasting moods.

Gargoyles perch by day on the rooftops,
With frozen stone smiles, and long, lolling mouths ajar,
No movement no sound 'til dark, and then GARGOYLE!
Zipping, slipping through the night air!
Laughing, leering tearing through the quiet air!
Scaring cats, chasing bats 'til dawn
When each one flits to his rooftop
And stares with his cold stone eyes at the busy streets
Through the only bright day,

---

[1] *Music Round the Town,* rev. by Max Krone *et al.* (Chicago: Follett Publishing Company, 1959), p. 78.
[2] Robert B. Smith and Charles Leonhard, *Discovering Music Together—Early Childhood* (Chicago: Follett Educational Corporation, 1968), p. 156.

And knows that night will come when GARGOYLE!
Laughing, leering tearing through the quiet air!
Scaring cats, chasing bats,
Reeling, spinning, shrieking, grinning, GARGOYLE! [3]

The quiet mood of the gargoyles during the day suggests a slow, deliberate tempo of reading. The wild nightlife of the gargoyles calls for a faster tempo. The capitalized word "GARGOYLE" signals the arrival of the exciting night for the gargoyles. You might try shouting this word to add emphasis to it.

Have children tape record themselves reading the poem individually. Also, several children may wish to read the poem together. You might ask them to listen for the different *qualities* in their voices. The light voices could read the descriptions of the daytime gargoyles while the darker voices could represent the nighttime gargoyles. When several persons read together, they are engaging in a choral speech activity.

After several days of orally interpreting the poem individually and in groups, the children will be ready to create a "sound picture" to accompany the reading. When developing the sounds to be used with "Gargoyle!" consider the *mood* established by the words of this poem. What kind of accompaniment do the words suggest? Will the same instruments be used to accompany the nighttime gargoyles as are used to accompany the daytime gargoyles? Perhaps rather than creating sound effects to enhance certain words of the poem, your class might develop an accompaniment using vocal, body, and instrumental sounds to establish the two contrasting moods of the poem. Using this idea, the children will create a mood environment through sounds.

Have several children read the daytime gargoyles' lines and several others read the nighttime gargoyles' lines while other members of the class produce the sound accompaniment for the reading. Invite the principal in to hear this performance. You might make a tape recording of the performance so that it could be played over the school intercom for the entire school to enjoy.

---

[3] "Gargoyle!" is found on pp. 107–108 of *Mastering Music* from the *New Dimensions in Music* Series published by the American Book Company. © 1970 by American Book Company. Used by permission.

The delightful poems about colors in *Hailstones and Halibut Bones* by Mary O'Neill are excellent sources to use in the exploration of creating a mood through sounds.

Using Tape-recorded Sounds

The children discovered that they could have interesting sound effects to accompany stories they read orally if they tape recorded the actual sounds. Some of the sounds they recorded were:

phone ringing
vacuum cleaner
walking on gravel
crash of waves against rocks
sharpening of a pencil
hitting a ball with a tennis racket
boiling water
popping corn
typewriter
breaking a branch
turning water on and off
letting air out of balloons

A recording of an actual rainstorm would point out that the falling of raindrops produces quite an irregular sound. Some raindrops seem to be louder than others. Why? You may wish to explore the concept of tempo in relation to the images created by the falling rain.

One class decided to write a story about their new school. They recorded sounds of the construction of a nearby house. They reasoned that the hammering and sawing of materials for the school would have sounded much like that of the construction of the house. They recorded sounds of the buses arriving in the school parking lot and of the children getting out of those buses. A great variety of sounds were recorded on the school playground. The children discovered that first graders' playground sounds are quite different from those of sixth graders. They recorded the morning announcements made over the intercom by the school's principal. Snatches of the day's lessons were recorded. These actual sounds from the school made the reading of their story about the new school more interesting than it would have been without them.

Collecting sounds on tape may be an interesting project for some members of your class.

One teacher accidentally played a tape that had been recorded at 3¾ speed at 7½ speed. The children were fascinated by the differences in sound that this mistake made. Some of the boys in this class were especially interested in manipulating the taped sounds they had collected. They played sounds that had been recorded at 3¾ at 7½ speed. They made a new tape of the original sounds played at 7½ speed. When they played the new tape at 3¾ speed, they discovered that it sounded very different from the original sound.

The music teacher had saved bits of broken tape. The children created interesting sound juxtapositions by simply using plastic tape to splice these bits together. Then, the boys decided to cut the magnetic tapes they had recorded themselves in order to edit it. Again, they spliced the tape using plastic tape.

Another interesting project with tape involves making "tape loops." [4] Record about one minute of a sound pattern that you like. To play the tape loop, put it in the recorder head and place the back of the loop around a chair. In this way, the desired sound will keep repeating. The sound recorded on the tape loop can become the ostinato background for another composition.

Predictable and Unpredictable Sounds

Do you remember the sound made when you rode over the railroad tracks on your bicycle? What sound did you hear as your little brother munched potato chips? How does the rain sound when it falls on a tin pan? When you walk along a picket fence and run a stick over the boards, how does it sound?

These questions may be used to stimulate thoughts about sounds. These images all represent experiences in which sounds play an important part. The sound of munching potato chips is unpredictable or irregular, whereas the sound made by running a stick over a picket fence is predictable or regular. What ideas can your class add to this list of predictable and unpredictable sounds?

---

[4] Another handbook in this series, *Electronic Music for Young People,* by Fred Willman, will provide you with further information on how to make a tape loop.

PREDICTABLE	UNPREDICTABLE
car turn signal	train whistle
tennis ball during game	typewriter
telephone ringing	faucet dripping
clock ticking	wind blowing

Perhaps your class would like to re-create these familiar sounds. Can they invent *signs* for each of the familiar sounds they list?

Photographs

Since so many children have cameras of their own, why not ask them to take snapshots of sound sources? These snapshots could serve as stimuli for the creation of an interesting composition. Ask the class to snap pictures of things that were making a sound as the photo was being taken. Some possibilities would include a dripping faucet, a truck dumping its load, a carpenter hammering, a motorboat zooming by, a rocket lifting off the launchpad. Using the photos in varying combinations could produce interesting sound patterns. The children could take a snapshot of something that ordinarily doesn't make a sound followed by altering the object in some way so that it does make a sound. The activity can result in a discussion of how sounds are produced. (See "Causes of Sound," page 9.)

Silent Sounds

Have you ever thought about the sound that a butterfly makes as it emerges from its cocoon? What about the sound of ice cream as it melts? Does a spider make a sound as it spins its web? Can you hear your fingernails grow? Will a falling star make a sound? When you peel a banana, do you hear a sound? Have you ever heard a sunrise?

None of these things produce sounds, yet each idea creates a particular image in the mind of a child. Your class may experiment with determining sounds that could appropriately represent images such as these, or others you think of yourself. For a moment, consider ice cream melting. *If* you could hear the melting of ice cream, would it be a harsh sound? a soft sound? What type of classroom instrument would be appropriate for representing this image? This type of activity may encourage your class to make musical decisions concerning the appropriateness of a particular choice of instrument on which to represent an image. *Decisions based on musical reasons reflect the emerging musical sensitivity of the children making the decisions.*

Sound Layering

The idea of placing one sound on top of another—or blending sounds together to form particular musical textures—may be explored using thoughts about everyday life as stimuli. You may wish to begin this activity by thinking about a single sound that is a part of a broader spectrum of related sounds. A honking car horn is a single sound; to this sound is added another honking horn and another. Soon a police car siren is heard. A dog begins to bark. Caution lights are flashing at regular intervals. Two cars crash together producing still another sound. The cacophony of honking horns, sirens and crashes develops the image of a crowded city street. The image of a crowded city street is built cumulatively by adding sound upon sound upon sound. The concepts of climax and tension and release can be explored and developed through activities such as this.

Following is a list of cumulative ideas:

—snicker-giggle-laugh-roar
—shades of red
—seed-seedling-small bush-sapling-full-grown tree
—a drop of water-rain-stream-river-waterfall

Fourth-grade children developed a composition based on the idea of fire. The children first discussed fire. They chose *words* that they felt symbolized the concept of fire. Then they discovered sounds to give life to their words. This experience was an exploration of *sound effects*. The children literally re-created sounds that they had previously heard. The composition began with the sound of the striking of a single match—lighting of a burner on a gas stove—explosion of grease catching fire—crackling of curtains on fire—roar of the house afire—wind whipping the fire through the woods. (This certainly was the climax of their cumulative buildup of sounds!) Release from the tension created was afforded by the coming of a rainstorm—sizzle of burning branches in water—whooooosh of water from the fire hose smothering the roar of the fire.

What instruments would you choose to create the sound effects listed by the children?

Your class may wish to explore some other sound effects:

—vacuum cleaner
—jet plane flying overhead
—clothes dryer with tennis shoes in it
—typewriter

Encourage your class to focus their ears to follow separate layers of sound. You might use *footsteps* as the sound source. Allow four students in different areas of the room to walk back and forth as you point to one or two of them. With their eyes closed, the other children should point to the direction from which the steps came. When two of the children walk at the same time, the listening students should point to two directions at once.[5] This experience could help your students to become aware of how sounds are put together. Can your students think of ways to symbolize this experience?

Sample Experience with Imitating Sounds

Objective: To select and combine sounds for the purpose of expressing through sound the images seen in a picture or photograph

Materials: A rhythm or environmental instrument for each member of the class
Pictures of your choice, which might include:

—picture of spectators cheering at a baseball game
—picture of people playing in the sand at the beach (waves should be seen in the background of the picture)
—picture of four children eating melting ice cream cones
—picture of a "haunted house"
—picture of a heavy rainstorm
—picture of carpenters building a house
—picture of a grandmother sitting in her rocking chair and knitting

Procedures:
1. Show the children in the class the pictures that you have gathered. Ask them to discuss what is happening in each picture. In your discussion,

---

5 James A. Standifer and Barbara Reeder, *Source Book of African and Afro-American Materials for Music Educators,* Contemporary Music Project, Music Educators National Conference, 1972, pp. 26–27.

emphasize the *sounds* that might accompany the activities in the pictures.

2. Divide the class into small groups of four to five children. Give each group one of the pictures that has been discussed. Tell each group not to show their picture to other groups.

3. Allow groups five to seven minutes in which they can select and combine sounds to suggest the meaning of the picture.

4. As each group performs its creation for the class, record the sounds. Have the children identify the picture for which the accompaniment just heard was created.

5. Give each group the recording of its composition. Each group should discuss how they might change their composition so that it would better represent their picture.

3

Composing

In this section, some suggestions for beginning compositions by the children will be given. You may want to add your own ideas as well as those offered by your students.

When beginning to experiment with musical composition, children need some specific suggestions and limitations as a framework within which to create. The structure and goal orientation provided by the teacher will aid students in making worthwhile creations.

Experimentation with environmental sounds, sound effects, and tape recordings provides children with a background for composing music. After many experiences with simple improvisations, they usually want to remember their compositions by notating them in some way. However, when these initial creative experiences are accompanied by emphasis on correct notation, children sometimes become frustrated and their creativity thwarted. Early compositional experiences should deal with sound exploration and not with the traditional symbolic representations of sound.

Establishing a Framework for a Composition

Perhaps we rush headlong into the business of translating music symbols into sounds. *The focal point of early musical experiences should be on sound exploration and discovery.* When your teaching emphasis is shifted from translating the jungle of music notation to experimenting with the sound potentials of various sound sources, children are free to improvise and experiment.

Suggest that the children use the sound-producers they have collected in order to *improvise* a pleasing composition. You may wish to divide your class into groups of four or five children for group composition work. Each child within the group should be actively involved in both the performance and creation of the composition. The emphasis of the early compositional experiences should be on *sounds* and not on symbolizing these sounds in traditional ways. It may be advantageous to change the members of the groups from one compositional experience to the next so that one child doesn't begin to dominate a group.

The way that a student chooses to work with sounds depends upon his previous experiences. The child who has had no previous experience should be allowed to experiment with both vocal and instrumental sounds so that he can discover their potentials.

A *music center* in a corner of your classroom might be established. Include in the center a variety of sound sources that have been collected by you and the children. At various times during the day, individual children might go to the center to compose or to listen to earlier compositions. A tape recorder with headsets will allow the children to listen to their original compositions without disturbing others. Differently colored magic markers, pencils and pens will be helpful for children who would like to notate their compositions. Graph paper, construction paper, and notebook paper should be available in the music center.[1]

After a child has developed a storehouse of sound experiences, you might wish to encourage him to experiment with a single musical idea to discover many ways of changing it into something different. The *development* of a musical idea is important to musical form.

An idea is changed when it is turned upside down, when it is elongated, when it is abbreviated, or when it is reversed. Each of these possibilities suggests a "compositional device": the musical term for an idea that is turned upside down is "retrograde"; an elongated pattern is "augmented," and an abbreviated idea is a form of "diminution." Here is a "picture" of a musical idea:

---

[1] For further information on creating music centers, see Sally Monsour's handbook in this series, *Music in Open Education.*

This is the same idea in retrograde—backwards:

Here is the original idea in augmentation—elongated:

When the idea is shown in diminution—abbreviated—it is shorter than the original:

Suggestions for Compositions

Early group compositions may be simply a recitation of isolated sounds; that is, each child will simply play his sound when his turn comes along. The following list of suggested compositions may help children to *develop* their original ideas:

Prepare a composition using the environmental "instruments" that you have brought to class. The instruents should be played solo and in varying combinations during the composition. Be able to perform your composition for the class. Your composition should be one minute long.

Create a composition based on this rhythm pattern: ♩ ♫ ♫ Try to show at least four ways in which the pattern can be changed, but remain recognizable as the same pattern. Use body sounds (tongue click, hand clap, finger snap, thigh slap, foot stamp, and so forth) only. Limit: One minute (have one member of the group use a stopwatch).

Using your environmental "instruments," develop a composition in ABA form. Try to incorporate at least three different ways of playing each instrument into your composition. Limit: Two minutes.

At this point, you might try to encourage the children to think about what makes music interesting. The ensuing discussion may include the following points:

Music is interesting when you hear different instruments playing.
Music is interesting when some parts of it are loud and others are soft.
Music is interesting when there are some fast sounds and some slower ones.
Music is interesting when it has some surprises in it.

After this discussion, allow the children to go back into their groups to work on their original compositions in an effort to make them more interesting.

Using Recordings

A recording can be an effective stimulus for creativity. It provides an immediate sound source to which the listener can relate. Many contemporary compositions emphasize unique ways of using traditional orchestral instruments. Some contemporary compositions incorporate environmental sounds into the music.

The Banshee by Henry Cowell uses a conventional instrument in a most unconventional way. The composition is to be performed on the *inside* of a piano. Strings of the piano are plucked, struck, strummed, and rubbed. Ask the class to listen to this composition to try to identify the instrument they hear. Children frequently believe that the instrument they hear in *The Banshee* is a member of the string family. Ask them *why* the composition could not be performed on only one stringed instrument (because the range of pitch variation in the composition is too large). Help them to discover that the piano is the instrument they hear. If you have a piano available for experimentation, the children will be very interested in looking at the inside of it and in exploring its sound possibilities. If you don't have a piano, an autoharp would be a good substitute. Encourage the children to experiment with the unconventional sound possibilities of the autoharp. Merely listening to and discussing Cowell's *The Banshee* will provide the children with many ideas of their own to explore.

If necessary, suggest the following possibilities for exploration of the autoharp's sound potential:

—Tap the autoharp strings with your hand, with a rhythm stick, and with a soft mallet.
—Explore the body of the autoharp for sound possibilities. Scrape a pencil over the chord bars, tap the sides of the instrument with a soft mallet.
—Tap several strings at once with the side of a pencil.
—Pluck individual strings of the autoharp to create a short, melodic ostinato.

Edgar Varese's composition *Ionization* was one of the pioneering efforts to mingle environmental sounds with traditional musical sounds.[1] Can you hear the airplane motor used in the composition? What other environmental sounds can you discover in the music?

Ben Johnston's *Knocking Piece* (1963) was written for two percussionists to perform on the inside of a grand piano.[2] According to Johnston, *Knocking Piece* is a "do-it-yourself" composition in which the composer has supplied the performer with a "kit" with which, within certain limits, he can work as he desires. Performance of *Knocking Piece* demands that one performer react to what the other is doing. The success of the performance depends upon the interaction

[1] *Ionization* and *The Banshee* are recorded by Folkways. *Sounds of New Music.* Album: Folkways Records FX 6160 © 1958. Source: Folkways/Scholastic Records, 906 Sylvan Ave., Englewood Cliffs, N. J. 07632.
[2] *Knocking Piece* is recorded on *Discovering Music Together—Early Childhood,* Listening Album L001, Follett Educational Corporation, Chicago, Illinois.

and relationship of the sounds produced by the two performers. Although the children won't be able to *hear* this interaction on the recording, it is an important point for you, the teacher, to recognize. You must encourage each child to listen to what he performs in relation to what other children in his group perform.

Knocking Piece achieves a great degree of *unity* because all of the sounds in it are, as the title indicates, knocking sounds. The children have already discovered that music is interesting when parts of it are alike (unity) and parts of it are surprising or different (contrast). Ask the children to discover what they hear in *Knocking Piece* that provides *contrast*. (Dynamic changes, various qualities of knocking sounds, pitch, rhythmic changes.)

In presenting *Knocking Piece* to the class, motivate the class by staging a "happening" in which two class members make various organized knocking (or other) sounds. This "happening" should be structured by the teacher and the two students before it is presented to the class. It should have a definite beginning and ending; pitch and dynamic changes should be evident.

Following the "happening," ask the students, "Does this 'happening' have anything to do with music?" Guide the discussion to include the following points:

—this was organized sound
—there was unity in the types of sounds heard
—a variety of timbres were heard
—a variety of dynamics were heard
—there were rhythmic patterns heard that could be notated

Then ask the students this question: "If you were to put together a piece of music using *only* rhythm, dynamics, and timbre, what problems would you have to solve?" Guide the discussion to include the following points:

—what types of sounds (timbres) will be used in the composition
—how to achieve unity and variety in the composition
—when each performer should play in the final composition; that is, the relationship of one performer to the others.

Following this discussion, play the recording of *Knocking Piece* for the students.

After the students have heard the recording, ask: "Is this music? What does *Knocking Piece* have in common with more traditional types of music? How could we define *music* on the basis of what we have heard today?" *Music is organized sound.*

As a result of this experience with *Knocking Piece,* the student should become more aware of the importance of unity and variety or contrast in composition. He can exemplify his awareness in his own composition.

This recording is an excellent source for stimulating students to create their own compositions using everyday sounds.

4

Vocal Sounds

Vocal sounds are among the most readily available sound sources in the classroom. The inflection, pitch, dynamic levels, and accent of spoken words all contribute to the meaning of vocal sounds. This section will give suggestions for using these sources in composing.

Discovering the Voice

The voice alone, without any spoken words, can convey feelings and meanings. A sigh is an audible sound that conveys a particular meaning without the use of spoken words. When an individual sighs, he is tired, bored, or a combination of the two. A moan is usually a low-pitched sound conveying the feeling of pain or discomfort.

Ask your class to discover other ways to use the voice to convey meaning without speaking. Have the children produce the sound rather than describe it. Can other children in the class describe the sounds that have been produced?

Make a list of the sounds made by the voice:

sighing	moaning
grunting	roaring
hissing	purring
cackling	heavy breathing
gasping	murmuring
whining	sneezing

Using Vocal Imagery to Free the Voice

Vocal sounds are among the most readily available sound sources in the classroom. Encourage the child to use his voice in unique ways by conjuring up images for him:

Many people don't like Mondays. If you were sad to see the weekend end and work beginning again, how would you say "Today is Monday"?

If you were excited about your weekend trip, how would you say, "Today is Friday"?

If your teacher said "Go straight to the principal's office!" how would you walk on your trip?

If Father gave you money for an ice cream sundae, how would you walk to the ice cream store?

In these experiences, children are exposed to concepts of mood, dynamics, tempo, and pitch.

You may wish to have the children experiment with the many possible ways of using (saying) a single word. For this experience, use an active word such as "thunder," "Halloween," "popcorn." By choosing only one word or concept with which to work, you provide subject matter for experimentation. Thus, the children will have to experiment with changes in articulation (staccato and legato), timbre, pitch, dynamics, texture, tempo, and rhythm, in order to create interest in the work.

Creating Vocal Ostinati

Examples from popular, folk, or classical music may help your children to focus on ways of combining sounds in unique ways, "2 – 4 – 6 – 8 —Who do you appreciate?" is an ostinato [1] used by the Jackson Five in their song "2–4–6–8." Encourage the children to create an ostinato pattern, either vocal, instrumental, or a combination of the two, as an accompaniment figure for their composition. An ostinato figure throughout the composition will lend a feeling of unity and coherence to a composition. After establishing the ostinato, the children can take turns improvising over the pattern. They may wish to form a vocal "orchestra" in which all those creating pinched-nose sounds sit together, all children making clenched-teeth sounds are grouped, all those performing whispered sounds are together, and so forth. This arrangement would allow a group of students to improvise together over the ostinato. Also, this arrangement would encourage the children to group *like* sounds.

To help the children establish a feeling of unity in their small group compositions, you may suggest that they work with a single idea

---

[1] Ostinato: a short, repeated pattern used as accompaniment; an ostinato may be spoken, sung, or played on an instrument.

of their own choosing. One group of students made an effective composition based on the topic "books." This ostinato pattern was:

"Books

                                             Books"
                           Books
            Books

Over this pattern, they chanted descriptive phrases about books: "library books, math books, hard books, easy books, red books, music
books. . . ." Each descriptive phrase was spoken at a different pitch
level, with varying dynamics and articulations. The tempo constantly
grew faster throughout the composition. The composition ended with
the entire group shouting the word "BOOKS."

A more experienced group of children created a short composition using "Bugs   Bugs   Bugs   Bugs" stated at steady intervals of
time as an ostinato pattern. To this, they added another ostinato
"Bee   Butterfly   Beetle," which extended over a two-beat period
of time. They then improvised "bug sounds" such as humming and
buzzing over their ostinato patterns. A class discussion following this
presentation pointed out the greater variety of sound material used in
the "Bugs" composition compared to the "Books" composition. In
their work, these children had incorporated many musical concepts
that had been discussed in class. By encouraging and accepting the
children's own ideas in the beginning, you are building a storehouse
of successful compositional experiences that can be refined as the
children have continued exposure to composing. Other children in the
class created compositions based on the following topics:

Candy bars: Mounds, Butterfinger, chocolate, red hot, peppermint
Pies: huckleberry, apple, blueberry, pumpkin, mint
Spring: rain falling, spshshshshshsh, birds chirping (effect), flowers
Cities: Boston, Chicago, New Smyrna Beach, New York
Shapes: f a t, skinny, short, tall

As children experiment with various sound possibilities, they
begin to discover that one's suggestion is better than another's because
it provides a contrast to the preceding idea. As they listen to each other
perform, the children discover that their vocal compositions can become more interesting when they are accompanied by instruments.
They may decide that the triangle is a better choice for a particular

section than is the woodblock because of the triangle's delicate timbre. This experience may be the genesis of the discovery of compositional devices such as variety, appropriateness, repetition, and so forth. *Such discoveries should be the emphasis of creative experiences. In these activities, the teaching emphasis should be on the creative process rather than on the finished product.*

Sample Experience with Vocal Sounds

Objective: To broaden the concept of the use of the voice for expressive purposes

Procedures:
1. Ask each student in the class to think of a short sentence to recite to the class.
2. Ask the students how their sentences would be changed if the following alterations were made:
   a. Make each sound in the sentence very short.
   b. Make each sound in the sentence very long.
   c. Make the first sound short, the second long, the third short, and so on.
   d. Make the pitch of each word in the sentence different.
   e. Begin the sentence very softly and gradually get very loud.
   f. Begin the sentence very loudly and gradually get very soft.
   g. Other.
3. Discuss the effect of the preceding changes on the meaning of the sentence spoken by each student.

5

The Look of Sound

What does sound look like? There are many ways of recording sounds. Symbolizing sounds by various marks on a piece of paper sometimes allows us to reproduce the original sounds at a later time. Graphing sounds by means of electrical instruments allows us to examine the unique characteristics of sounds produced by a variety of sources. Records and tape recordings allow us to capture sounds exactly as they occur so that we may listen to them at will.

You can "see" sound when you immerse a ringing tuning fork into a pail of water. The sound vibrations send waves out from the point of immersion. Another way in which you can see sound vibrations is to put sand on a drumhead and then tap the drumhead. The sand will "jump."

The Oscilloscope

We know that the sound of a flute playing "A" is different from the sound of a piano playing "A," although we recognize that both instruments are playing the same fundamental tone. Why do these two instruments sound different? An oscilloscope could help us to *see* the answer to this question. The oscilloscope is an instrument that registers the variations of an electrical current and photographically records them. When the sound waves that are produced by each musical instrument are "photographed" by an oscilloscope, the "picture" obtained for each instrument is different. Each picture maintains the general shape of the fundamental tone, but has a distinctive pattern of extra "squiggles" caused by the overtones or harmonics of the particular instrument.

Voice Prints

Voice prints (Figure 5-1) that *show* the distinctive features of an individual's voice are now being developed. Each person's voice has

37

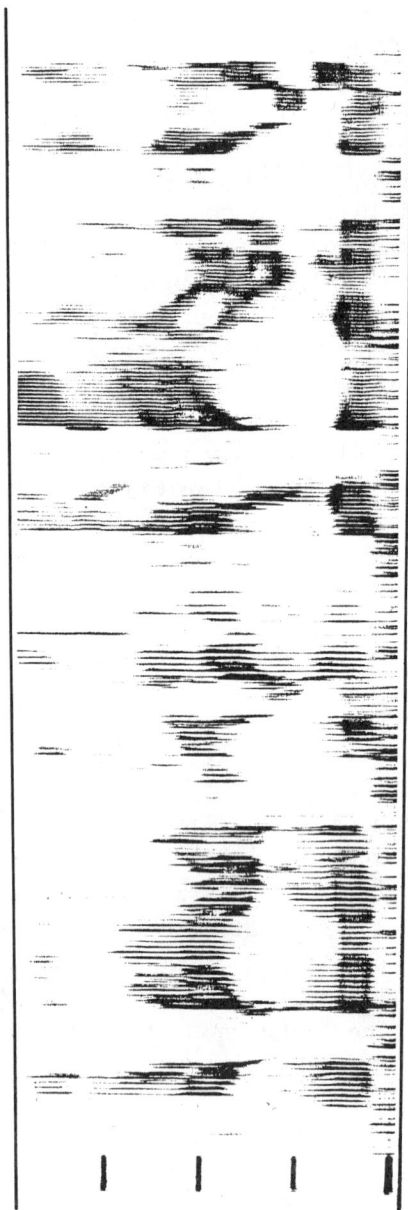

Figure 5-1

a unique sound; consequently, the "print" of his voice is different from the print of anyone else's voice. Someday, voice prints may be used to identify an individual in much the same way as fingerprints are used today.

Drawing on Magnetic Tape

By drawing on magnetic tape (the kind used for a reel-to-reel tape recorder), you can graphically create a pattern of sound. Straight lines etched into the tape produce a particular sound, while holes poked into the tape produce another. This may be an interesting project for your class.

Music Symbols

Two main reasons for musical notation suggest themselves:

1. to help the creator *remember* his composition, and
2. to allow the creator to *share* his composition with others— whether they be from the next room or the next century.

Beginning Compositional Symbolizations

The need for written symbolization of music wasn't felt until the end of the sixth century. By that time, there were far too many Gregorian chants (plainsongs) for anyone to remember. The church's first experimental music notation consisted of a series of short curves that followed the rise and fall of voices. These curves were called *neumes,* the Greek word for signs. The neumes were sufficient to remind the church singers of the tunes that they had to sing, but they were not *exact* enough to enable a singer to learn a new plainsong melody by reading them.

By the end of the eleventh century, a horizontal line running from the beginning to the end of the chant was added. The line had a letter against it to indicate its pitch. This lettered line enabled the singers to determine the pitches of all of the other lines and spaces of the staff. By this time, musicians used written "notes" instead of the neumes. The musical notation used in the twelfth century enabled cathedral choirs to know the exact pitches on which they were to sing, as well as the relative lengths of those pitches.[1]

Another early form of musical notation was the *tablature*. The tablature showed a "table" of the notes to be played. It told the player *what to do* (which finger to use and where to put it on the string or keyboard), whereas other forms of notation showed the *sounds to be produced*. Today, tablatures are still used to notate ukulele music, for example. The tablature enables the person who is unfamiliar with traditional musical notation to perform on a musical instrument.

Twentieth-century compositional practices are becoming so complex that traditional means of music notation cannot always accommodate the changes. Some sounds simply cannot be shown using traditional means. The staff provides no way to deal with microtonal compositions, for example. Old types of notation frequently don't work for composing electronic music; symbolization for electronic music frequently takes the form of graphs. A music synthesizer is capable of producing sounds that humans are incapable of producing. Quick changes from one sound to another can be produced by programming a computer to make the changes. Directions for the computer must be written in computer language rather than in music notation.

Graphic Notation

Many composers of contemporary serious music use "graphic notation." Graphic notation is designed to stimulate a performer's imagination. *Diversions for Piano and Clarinet* by David Eddleman [2] uses graphic notation (Figure 5-2). Through the symbols he has chosen, Eddleman attempts to convey a concrete image of the sounds he has assembled in this composition. The line Eddleman calls "sys-

---

[1] Some children in your class may wish to explore early forms of musical notation. They may begin by referring to *The Wonderful World of Music* by Benjamin Britten and Imogen Holst (Garden City, N.Y.: Doubleday & Co., 1958) and *The Oxford Junior Companion to Music,* by Percy A. Scholes (New York: Oxford University Press, 1959).

[2] *Diversions for Piano and Clarinet* is available from Media Press, Box 895, Champaign, Ill. 61820. Order No. MP 2203–2.

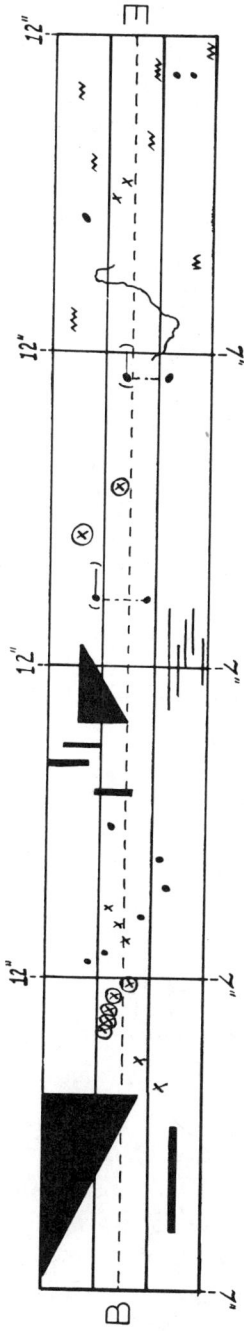

Figure 5-2

41

tem B" is divided into three sections; these three sections represent the three parts, low, middle, and high, of the piano keyboard and of the clarinet range. The first symbol encountered in system B is this

 . The symbol itself indicates that the system begins with

a high sound and gradually additional sounds are added until the sound spans from the high to the lower middle range of the piano. At the end of the first "measure" of the system, Eddleman uses these

symbols . These indicate separate sounds that are played

in the middle range of the piano.

A key to the symbols is included with the score of this composition. The symbols that the composer uses here are meaningful to him, but in order for a performer to interpret the symbols in the same manner, Eddleman must specify his meanings. The following list indicates the composer's interpretation of his symbols:

 A tone cluster, the size depends on the thickness of the symbol.

In traditional notation a tone cluster may look like this:

 A crush cluster in which notes are gradually added.

In traditional notation, a crush cluster may appear:

A single note played staccato

A single sustained tone

Pluck inside strings with damper raised

In *Diversions for Piano and Clarinet*, the manner in which the instruments are played is unusual. There are no traditional music symbols to indicate this composer's intentions. Consequently, he was forced to create symbols that would serve his own purposes.

When a performer is faced with unfamiliar notation that only outlines the material he is to perform, his attention is shifted from interpreting symbols to the stuff of music itself—*sound*. The directions and suggestions given by the composer provide the performer with a general outline of the composition without dictating to him exactly how and what he is to perform. The performer is left to make musical decisions of his own. He is left to experiment with sound and with the relationship of his sounds to those of other performers in his group. The performer is *involved* in the production of music on a truly creative level.

Another advantage of using graphic or nontraditional means of notating sound is that *all* students can participate in the creative activities. Performance of music is not limited to those who can read music when graphic notation is used.

Larry Phifer's *Construction No. 2* [3] (Figure 5-3) shows traditional music symbols in very untraditional arrangements. The composition is to be played by any combination of three to six instruments including at least one percussionist. The fact that the composer has indicated this emphasized that his interest lies in experimenting with various sound possibilities, rather than with strictly imposing his musical choices on the performers. The performers are given further freedom to experiment to determine how they wish to have the music sound. The instructions are that the performers may begin in any parallelogram on the score. The movement from one parallelogram to another must follow the connecting lines, although repetition or reversal of a parallelogram is allowed by the composer. Again, the performer is encouraged to experiment with the sound possibilities within the framework given by the composer.

English composers George Self, Brian Dennis, David Bedford, and others concerned with classroom uses of music have adopted a simplified system of graphic notation. The unconventional symbols used in their notation are used by many English composers. The symbols are easily understood even by the individual who has no knowledge of conventional musical notation.

Bernard Rands has created four different "Sound Patterns" utilizing this simplified graphic notation.[4] The Sound Patterns are

---

[3] *Construction No. 2* is also available from Media Press, Box 895, Champaign, Ill. 61820. Order No. MP 1704–2.

[4] "Sound Patterns" are available from Theodore Presser, Presser Place, Bryn Mawr, Pennsylvania 19010.

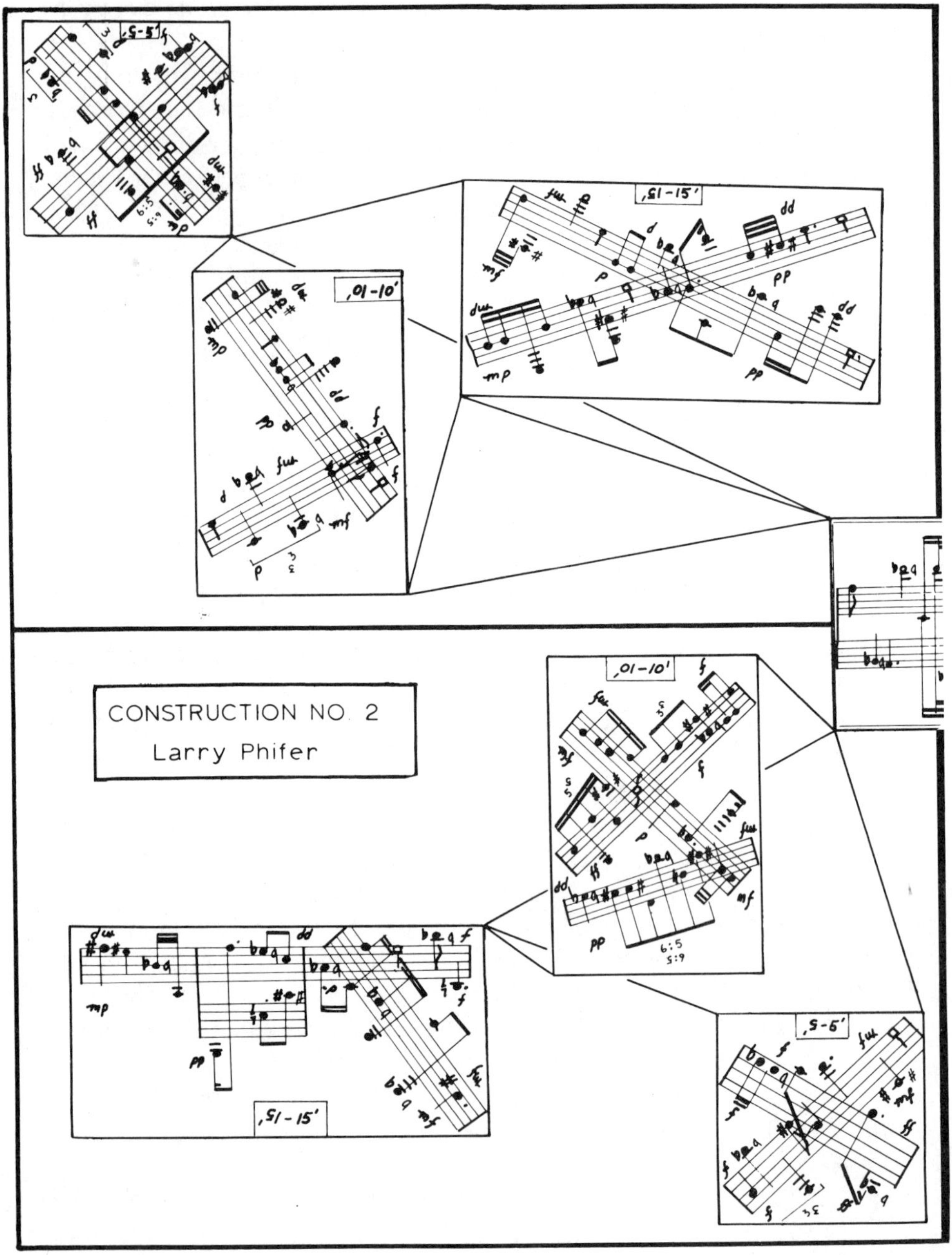

Figure 5-3

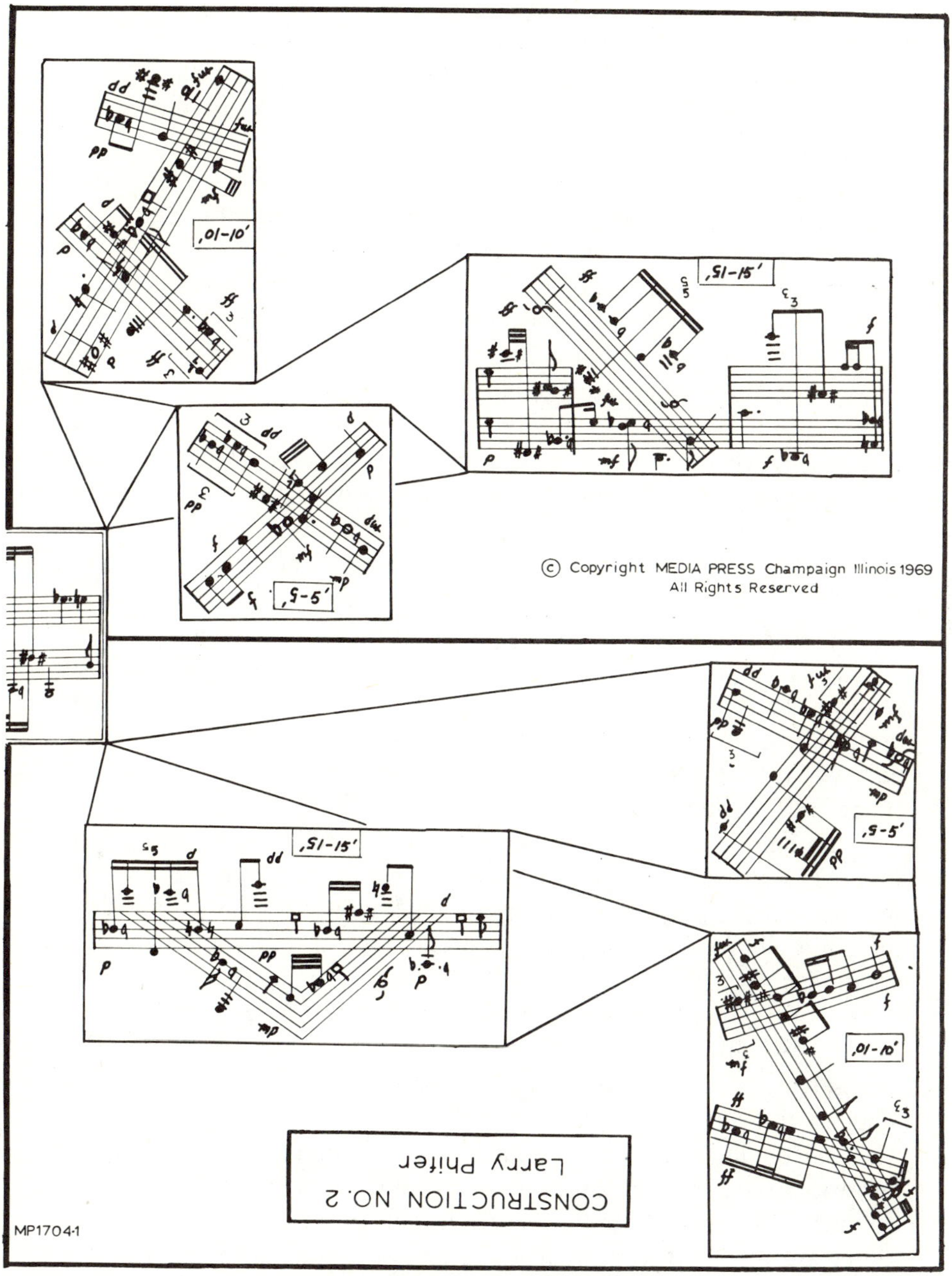
© Copyright MEDIA PRESS Champaign Illinois 1969
All Rights Reserved
CONSTRUCTION NO. 2
Larry Phifer
MP1704-1

Figure 5-3 (cont.)

written for voices and hands. When "Sound Patterns 1 for Voices and Hands" was presented to a group of fourth graders, they responded readily to the new notational system. On the day following the presentation, the children were able to create individual compositions utilizing these graphic symbols.

For some children graphic notation may provide an intermediate step between improvising and composing using traditional notation. Children may wish to write the results of their experimentations with sounds if they are encouraged to develop symbols to which they can relate. Musical creativity stems from becoming sensitive to sounds themselves and from manipulating sounds in meaningful ways, and not necessarily from correctly interpreting music symbols.

Many contemporary composers advocate an empirical method of composition in which the composer goes directly to the sound materials and experiments and improvises with these until he has fashioned a coherent sound structure. Electronic music affords the composer the opportunity to *hear* his music performed as he is creating it. Using a music synthesizer, the composer works directly with sound materials rather than with symbols to represent those sounds.

Encourage the children in your class to work directly with sounds—to improvise. Ask them to *remember* what they have improvised so that they can repeat and develop their improvisations. In this way, the improvisor can work with his sound material until the shapes of his sound patterns are satisfying to him.

To help children remember their compositions, you may suggest that they write or draw their ideas in a way that helps them recall what they wanted to play. By making a visual symbol of his composition, the child can *look* at what he has created in an effort to improve his creation.

Early symbolization of musical ideas may be as simple as that shown in Figure 5-4. Even before beginning to interpret the symbols themselves, the performer is faced with the problem of determining whether to read the graph from left to right or from top to bottom. This particular score·is intended for one performer to read and perform from left to right. Scoring becomes more difficult when two or more instruments are involved in the composition. Most often, scores are intended to be read from left to right. However, when two or more instruments are playing together, the performers are required to read up and down (to determine the relationship of one part to

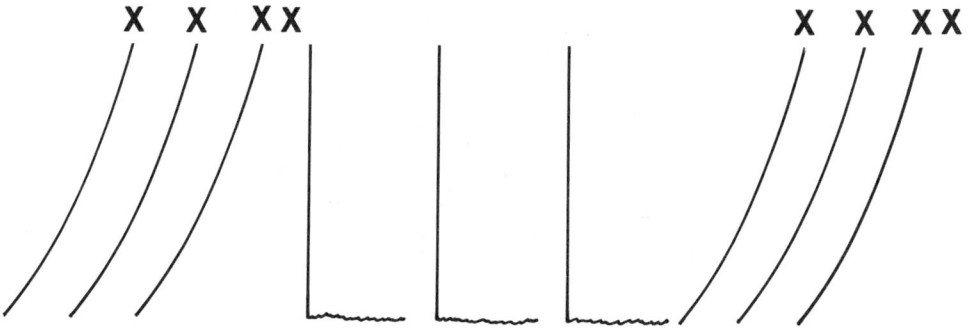

Figure 5-4

another) as well as from left to right. Students must consider this problem when they are notating compositions for more than one performer.

For the foregoing composition, a road-map type of key to the symbols helps a new performer to understand the composer's intentions. The composition shown above was written for notched rhythm sticks.

 —rub sticks together in an upward motion

X —hit sticks together

 —drop stick and allow it to bounce

Although the composer of this work intended it for the notched rhythm sticks, could it be played on any other instrument? Your class might experiment with playing it on the piano. When this composition is played on the piano, the symbols might be interpreted thus:

 —upward glissando or arm roll moving upward on keyboard

X —high-pitched tone cluster

 —downward glissando followed by arm roll in lower register

The children in Mrs. C's class were so excited about the composition they had created that they wanted to share it with another class. After they had decided exactly how the composition should sound, they had several rehearsals to polish their performance. Finally they were ready to perform. At a crucial point in the final performance, Jamey forgot his part! The other children were very upset because they felt that this memory lapse had ruined their performance.

This incident pointed out the need to have signs or signals to tell performers when to play. The children recalled that they always saw a conductor directing the music at the local orchestra performances. They felt that a conductor would be a great help in telling performers when to play their parts. Hillary asked, "What if the conductor forgets whose turn it is to perform?" A discussion of this very real problem lead the children to determine that some of the memory problem could be alleviated if they wrote down the sequence of sounds in their composition. The example in Figure 5-5 shows their composition, "Flowers."

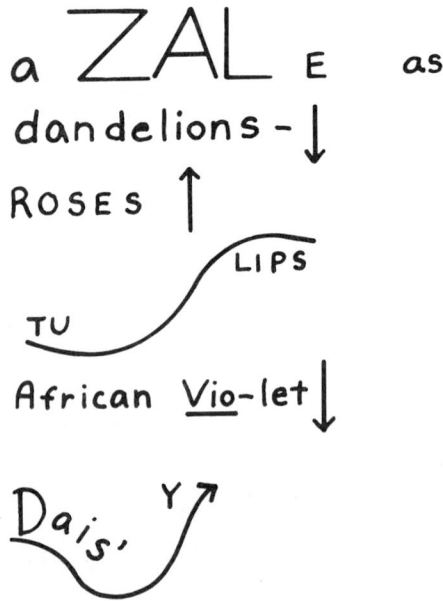

Figure 5-5

The notation for "Flowers" was created by the children. The capital letters indicate words or parts of words that are to be spoken with a loud voice. The arrows indicate the pitch of the words; notice the sliding pitch of the word "daisy."

After writing down a composition, it becomes more *concrete*—you can *see* what you have created and you can re-create the composition in the same way as it was planned. The written version of a composition provides a means for allowing more manipulation of the sound material in the composition because the composer can *see* what he has created.

At this point, you will need to provide some guidance for the children's exploration. You can *focus* the initial period of notational exploration by helping the children to recall compositional techniques discussed and experienced earlier. They may recall the following ideas for creating interest in music:

Change rhythm
Turn shape upside down
Alter articulation
Reflect sound by resonating it (for example, play a melodic pattern on the temple blocks and repeat the pattern on the glockenspiel or resonator bells)
Play an idea backwards (retrograde)
Play an idea twice as fast as usual (diminution)
Elongate an idea (augmentation)
Repeat an idea at different pitches
Change the linear order of ideas
Rearrange the vertical order of materials

The emphasis of the compositional experience should be on what is behind *the symbols—what sounds are created by means of the symbols.*

The child should be encouraged to experiment until he finds sound patterns with a satisfying shape. This experimentation by individuals could take place in the music center of your room. The shape of the composition shown in Figure 5-6 satisfied Diana. Her composition had a definite direction. The basic idea of the composition was repeated and changed in a variety of ways; the repetitions and contrasts afforded in the composition provided a feeling of wholeness.

After your class has had varied experiences with manipulating and extending sound materials, you might discuss the concept of musical form or shape with them. *The form of a composition is the result of the repetitions and contrasts within the sound structure.* A composition that is merely the juxtaposition of a vast sum of sound materials fails to be satisfying because it provides no development of

Figure 5-6

material. The listener needs to hear the same material repeated in a composition so that he has something to "grasp onto"—something that gives direction to his listening.

You may suggest that the children explore ways of combining sounds.

Teacher: Look at the sentence I wrote on the board.
Class: I like to hear the birds singing.
Teacher: Good! I have written the sentence over here in a different way.
I like to hear the birds singing.

                I like to hear the birds singing.
Let's divide into two groups; group one will begin by reading the top line. Group two will begin to read the second line when I point to them. Ready?
Class: I like to hear the birds singing.

                I like to hear the birds singing.
Teacher: What kind of a song do we sing in which we divide into groups to all sing the same song but begin singing at different times?
Gay: That's a round!
Teacher: Right! Look at how our words are written over here.

I like to hear the birds singing.

I like to hear the birds singing.

I like to hear the birds singing.

I like to hear the birds singing.

I like to hear the birds singing.

How do you suppose that we should say the words this time?

Mary: I think that we should have someone stand in front and say the sentence and someone stand on the side of the room by the fishbowl to say the sentence. Someone else should stand in the back of the room and a fourth person should stand by the light switch to say it. The one in the front says it first and then the person by the fishbowl and then in the back and then by the switch.

Teacher: That sounds like a very good plan. Boys and girls, why do you think Mary's idea is a good one?

Ginger: Because that's how it looks on the board. The arrows show us which way the sound moves around.

Teacher: Yes, that's a very fine explanation. Let's perform it.

The class listened as the sound of the spoken sentence moved around the room.

Teacher: Look at how the sentence is written here.

I like to hear the birds singing.
 I like to hear the birds singing.
  I like to hear the birds singing.
   I like to hear the birds singing.

I like to hear the birds singing.
I like to hear the birds singing.
I like to hear the birds singing.
I like to hear the birds singing.

How many groups will we need to perform this time?

Bob: We need four groups because there are four lines.

Tom: I think we need eight groups because there are two sets of four lines.

Sue: And I think that one set of four people should stand by the light switch and the other set of four should stand by the fishbowl.

The class listened to the different effect of having words overlapping and coming from different parts of the room.

Teacher: That was very interesting! How would you read the sentence when it's written like this?

I LIKE to hear the birds singing.

Bill: I'd shout LIKE.

Teacher: Good idea. Let's hear you try it.

The class then explored the possibilities of emphasizing different words by shouting them.

Teacher: How would you read the sentence when it's written like this?

I like to hear the birds singing?

The inflection of Ron's voice as he read the sentence indicated that the declarative sentence had now become a question.

Teacher: How would you read the sentence when it's written like this?

I LIKE

                                      singing.

              to hear the birds

Peter's rendition of this sentence showed differences in pitch.

Teacher: Very good. What can you tell about how the way words look affecting the way in which they are read?

Bonnie: When sentences are divided up so some words are high on the page and others are low, then you know to say some with a high pitch.

Tom: When some words are written in capital letters, they are louder than the rest.

Carol: Arrows can tell you which way to read the sounds.

Teacher: You have all given good answers. Now let's divide into groups of five people. Each group should create a composition using a single sentence. First, agree upon the sentence that you will use. Then think about the interesting ways that you could change that sentence. Experiment with the sounds in the various ways you have talked about. After experimenting, write down some of your ideas. What you write down should *show* you *how* you have decided to perform the sentence. That way, when you work on the composition again tomorrow, you will be able to remember what you did today. We will work on this project for several days. Now you may choose your groups.

One group's composition used as its theme "I like to drink malts." Part of their composition played with echoing as a means of variety. The notation they used to show this way was simply:

I LIKE TO DRINK MALTS.

                                i like to drink malts. (echo)

Another part of the composition used "fragmentation." One word of the sentence was spoken by each of the five members of the group.

Joe: I
Jim: like
Marie: to
Rhonda: drink
Rick: malts.

Yet another section of their composition involved layers of sound. The children's interest in experimenting with colored markers on overhead projector transparencies lead them to make a series of transparencies showing their theme written in different ways. Jim's rendition looked like this:

Megan's transparency was done in red:

I like to drink malts.

David's work was done in very bold letters:

I like to drink malts.

Susan printed her words very delicately:

i like to drink malts (chocolate!)

Barbie's idea was different from the others:

like

I

drink

malts.

to

Jim put his transparency on the overhead projector first. He began to chant the theme in a way that he had reflected in his notation. Megan placed her transparency on top of Jim's and she began to make her sound pattern. David began his sound as he put his transparency over Megan's. Susan put her transparency under the entire stack of transparencies! Finally, Barbie had her turn to place her transparency on the others. The children decided that they had created an additive composition. Then, they began to subtract their sounds (and transparencies). Barbie was the first to remove her transparency, second David, then Megan, then Jim, and finally Susan. (See page 13 for the notation of a similar type of composition using environmental instruments.)

Spatial Arrangements

In this experience, the children were lead to discover the effect of spatial arrangements on musical performance. Is a different effect created when the same sounds are heard coming from different parts of the room? Does the same idea performed by different groups of students sound exactly the same? What effect does the proximity of the performer to the audience have on the perception of the sounds?

By altering the spatial relationship of one group or performer to another, you have produced variety in the composition. You will probably wish to explore the effect of various spatial arrangements upon musical performance.

This work may lead to the exploration of the theme and variations form. Encourage the children to develop an interesting theme of their own or to select the theme of a familiar song. Then, allow approximately fifteen minutes for groups of five to six children to create two or three variations on their theme. In addition to alterations of spatial arrangements, what types of variations might the children include?

—adding a chant, ostinato, or descant part to the theme
—changing the rhythm by augmentation, diminution, syncopation
—changing the mode (from major to minor, for example)
—using different instruments to play the theme
—using different instruments to play each note of the theme (fragmentation)
—varying the dynamics

The spatial arrangement of the traditional concert situation is of interest. The typical concert setup seats the orchestra or other performing group apart from the audience on a stage. Many contemporary composers feel that this separation of the performers from the audience is dangerous. Some have created compositions in which the audience actually performs. The involvement and participation of the audience in the creation of music can be a stimulating experience.

Pauline Oliveros' composition *Meditation on the Points of the Compass* is an audience-participation piece.[5] The composer's directions state: "This composition is conceived for a large open space. . . . Before the piece begins, the conductor verbally instructs the audience when and what sounds to make."

Small groups of children in your class may create compositions for audience participation. Encourage them to devise their own spatial arrangements.

---

[5] *Meditation on the Points of the Compass* is available from Media Press, Box 895, Champaign, Ill. 61820. Order No. MP 3105–1.

Transcriptions

After the children have explored the various combinations of vocal sounds and the notation of these, suggest that they transcribe the compositions they have written for voices into compositions for instruments. The children should determine which instruments would best portray the mood of their compositions. How could they utilize the instruments to create the effects that they produced with their voices earlier? Since the composition is based on a single sentence, should only one instrument be used throughout the composition? When the composition was spoken, did everyone's voice sound the same? Will the notation of the composition be the same for instruments as it was for voices? Children will have to answer these and other questions in order to be successful with this experience.

After the children have transcribed their compositions, allow each group the opportunity to tape record them. You may ask the groups to exchange the notation for their compositions. Allow each group to perform another's composition. You might then compare the performance by the group who created the composition with the group who simply re-created it. Were the performances the same? How were they different? What reasons can you give for the differences? This discussion may lead the children to discover that graphic notation may not be as precise as conventional notation. The use of imprecise symbols may cause the performers to become more involved in the re-creation (performance) of the composition because in order to perform it, they need to determine how it should sound.

Sample Experience with the Look of Sound

Objective: To create symbols to represent expressive sounds

Materials: Tape recordings of environmental and vocal sounds
Pencil and paper for each student
Each recording should emphasize a single element of music, e.g., pitch, dynamics, length of sound, pattern of sound, etc. Examples of some possible sounds are:

car horns honking
rain falling
voice sliding upward
voice sliding downward

single word spoken at different pitch levels

single word spoken in different lengths (i.e., sometimes very short and sometimes longer)

Procedure:

1. Give each student a large piece of paper and a pencil. Ask the students to listen to the tape recording of sounds you have recorded.
2. Play an example of the voice sliding upward and discuss the sound. Ask students to use lines, shapes, or anything else of their choice to show the direction of the sound heard.
3. Discuss the usefulness of various class members' symbols for the sound heard.
4. Play an example of a single word spoken at different pitch levels. Ask the students to use symbols of their choice to show what they heard in this example.
5. Discuss the second example. Students should discover that pitch was changed each time the word was repeated. Their pictures of the sound should show different pitches.
6. Play several more examples for the students to notate using symbols of their own choice.
7. Let each child work with a partner to create several sound patterns for other members of the class to notate.
8. Allow some pairs of children to play their sound patterns for other members of the class to notate.

6

Movement

The children have "planted their feet in cement" so that they can explore ways of moving their arms, heads, and torsos.

—"Pretend that your bodies are tiny feathers that are floating down to the ground. Start with your arms stretched up as high as you can reach.

—"Pretend that you are very angry at your pillow. Clench your fists and start hitting your pillow, which is lying right in front of you in the air.

—"Pretend that you are a twig that has just been caught in a waterfall. Suddenly, you are on the edge of the fall . . . and you slide right down to the bottom of the waterfall."

Almost anything done in movement can be translated into sound patterns. Working on different *levels* suggests different *pitches;* the *speed* of the movement dictates *tempo;* the *type* of movement suggests instrumental timbres, dynamics, and articulations.[1]

Changing Appropriate Instruments to Accompany Movement

When a gong is struck, the sound resonates and lingers in the air for a period of time. Such a sound may suggest a slow, sustained type of movement. "Which of the 'pretends' could be accompanied by the sound of the gong? Does the sound of a downward glissando played on the melody bells suggest a particular type of movement to you? A collapsing movement may correspond to the downward glissando. Do any of the 'pretends' indicate a collapse? The crisp sound of unregulated rhythm patterns played on the woodblock could result in a sharp, percussive type of movement. Which 'pretend' will the woodblock accompany?"

---

[1] For additional information see *Rhythms Today!,* by Edna Doll and Mary Jarman Nelson (Morristown, N.J.: Silver-Burdett Co., 1965).

59

Shades of a Color

Children are familiar with colors. Colors can be interesting stimuli for self-expression. Give each child a piece of differently colored paper. Ask each to write words he thinks of when he looks at his colored paper.

> Do you feel funny? sad? gay?
> Is the color an exciting one? a calm one?
> Does the color remind you of big? of little?

Ask the children to draw *lines* (not pictures, just lines) to describe the adjectives they have written. Allow the children an opportunity to experiment with movements they feel reflect these lines.

Different shades, intensities, and gradations of a particular color can extend the initial discoveries. Ask the children to find others who have a different shade of their color. You may have four or five shades of a single color. Comparisons of the adjectives and lines used to describe the varying shades of a particular color can be quite interesting. Each color group of four or five children could create its own movement patterns to represent the variations of its color using sound sources and combinations to best reflect the various shades of the color. You might suggest that each color group develop both movement and sound patterns to represent its color.

Using a continuum of reds ranging from fire-engine red to baby pink as stimuli, one group drew the lines shown in Figure 6-1. For the brightest red, they selected a percussive pattern played on the guiro. The next shade was symbolized by the sound of the cymbal clanging. The third gradation was played by scraping fingernails in

Figure 6-1

circles on a drumhead. The lightest shade of red was played on the finger cymbals. The movements that reflected fire-engine red were jagged arm slices performed with clenched fists. The children stood very rigidly, presumably to demonstrate intensity. The percussiveness of the movements diminished until finally the children used very gentle arm sways to represent baby pink. For this palest shade of red, they were in a kneeling position. Through their choices of instruments, movements, and levels of performing the movements, the children demonstrated a great deal of sensitivity to this project.

Multimedia Spectacle

Have you ever made a movie without using a camera? Secure used 16 mm film from a nearby television studio or from your media center. Remove the pictures that have been recorded on the film by dipping the film in bleach. Wet cotton with bleach and rub it over both sides of the film. This removes the emulsion, leaving the film clear. After this process, you are ready to create a movie without a camera.

The children can draw whatever they wish on the film with magic markers (the type used to write on transparencies work well). Remind them that each frame is only projected for one second. Consequently, if they want a particular part of their drawing to appear for a period of time, they must draw it exactly the same many times.

When the film is completed, encourage the children to create a sound accompaniment for it. The result of this project can be an exciting combination of art, music, and literature skills and it may provide the highlight of a P.T.A. meeting!

Music must be heard, not just seen or talked about. The children in your class will probably want to perform their compositions for others—mothers, friends, a local T.V. or radio station, or the P.T.A. The audience will find it interesting to see the notation of the compositions created by the children as an accompaniment to the music. The notation can be traced from the original onto a transparency projected by an overhead projector.

Sources for Further Exploration

The following list of books are useful sources for additional exploration of the ideas presented in this handbook.

Brian, Dennis, *Experimental Music in Schools Towards a New World of Sound*. London: Oxford University Press, 1970.

Clemens, Gwen, *Making Music* (Children's Book). London and Harlow: Longmans Green and Co., Ltd., 1966.

——————, *Making Sounds* (Children's Book). London and Harlow: Longmans Green and Co., Ltd., 1966.

Marsh, Mary Val, *Explore and Discover Music*. Toronto: Collier-Macmillan Canada, Ltd., 1970.

Paynter, John, *Hear and Now: An Introduction to Modern Music in Schools*. London: Universal Edition, 1972.

——————, and Perter Ashton, *Sound and Silence: Classroom Projects in Creative Music*. London: Cambridge University Press, 1970.

Self, George, *New Sounds in Class: A Contemporary Approach to Music*. London: Universal Edition, 1967.

The following articles, all from the *Music Educators Journal,* will be of interest to you:

Boardman, Eunice, "New Sounds in the Classroom" (November 1968), 55:3, pp. 62–65.

Elliott, Dorothy Gail, "Junk Music" (January 1972), 58:5, pp. 58–59.

"The Handwritten Sound Track" (March 1968), 55:3, p. 114.

Hoenack, Peg, "Unleash Creativity—Let Them Improvise!" (May 1971), 57:9, pp. 33–35.

Holderried, Elizabeth Swist, "Creativity in My Classroom" (March 1969), 55:7, pp. 37–39.

Horazak, Thomas J., "Focus on the Unconventional" (October 1973), 60:2, pp. 49–51.

Johnson, Tom, "Teachers, Step Up to the Avant-Garde!" (May 1972), 58:9, pp. 30–32.

"Original Student Compositions" (March 1968), 55:3, pp. 83–85.

Reese, Sam, "New Music Breeds Creators, Not Repeaters" (January 1973), 59:5, pp. 65–67+.

Tait, Malcolm J., "Involving the Young Child in Music: Whispers, Growls, Screams and Puffs . . . Lead to Composition" (February 1971), 57:6, pp. 33–34.